CHEMISTRY RESEARCH AND APPLICATIONS

THE PROMOTING EFFECT OF LANTHANUM IN HETEROGENEOUS CATALYSTS

CHEMISTRY RESEARCH
AND APPLICATIONS

Additional books in this series can be found on Nova's website
under the Series tab.

Additional E-books in this series can be found on Nova's website
under the E-book tab.

CHEMISTRY RESEARCH AND APPLICATIONS

THE PROMOTING EFFECT OF LANTHANUM IN HETEROGENEOUS CATALYSTS

MARIA DO CARMO RANGEL
MANUELA DE SANTANA SANTOS
AND
SERGIO GUSTAVO MARCHETTI

Nova Science Publishers, Inc.
New York

Copyright © 2011 by Nova Science Publishers, Inc.

All rights reserved. No part of this book may be reproduced, stored in a retrieval system or transmitted in any form or by any means: electronic, electrostatic, magnetic, tape, mechanical photocopying, recording or otherwise without the written permission of the Publisher.

For permission to use material from this book please contact us:
Telephone 631-231-7269; Fax 631-231-8175
Web Site: http://www.novapublishers.com

NOTICE TO THE READER

The Publisher has taken reasonable care in the preparation of this book, but makes no expressed or implied warranty of any kind and assumes no responsibility for any errors or omissions. No liability is assumed for incidental or consequential damages in connection with or arising out of information contained in this book. The Publisher shall not be liable for any special, consequential, or exemplary damages resulting, in whole or in part, from the readers' use of, or reliance upon, this material. Any parts of this book based on government reports are so indicated and copyright is claimed for those parts to the extent applicable to compilations of such works.

Independent verification should be sought for any data, advice or recommendations contained in this book. In addition, no responsibility is assumed by the publisher for any injury and/or damage to persons or property arising from any methods, products, instructions, ideas or otherwise contained in this publication.

This publication is designed to provide accurate and authoritative information with regard to the subject matter covered herein. It is sold with the clear understanding that the Publisher is not engaged in rendering legal or any other professional services. If legal or any other expert assistance is required, the services of a competent person should be sought. FROM A DECLARATION OF PARTICIPANTS JOINTLY ADOPTED BY A COMMITTEE OF THE AMERICAN BAR ASSOCIATION AND A COMMITTEE OF PUBLISHERS.

Additional color graphics may be available in the e-book version of this book.

LIBRARY OF CONGRESS CATALOGING-IN-PUBLICATION DATA
Rangel, Maria do Carmo.
  The promoting effect of lanthanum in heterogeneous catalysts / Maria do Carmo Rangel, Manuela de Santana Santos, and Sergio Gustavo Marchetti.
    p. cm.
  Includes index.
  ISBN 978-1-61761-272-5 (softcover)
  1. Lanthanum. 2. Heterogeneous catalysis. I. Santos, Manuela de Santana. II. Marchetti, Sergio Gustavo. III. Title.
  QD181.L2R36 2010
  546'.411--dc22
                                      2010031452

Published by Nova Science Publishers, Inc. † New York

CONTENTS

Preface		**vii**
Chapter 1	Introduction	**1**
Chapter 2	Applications of Lanthana as Catalysts and Supports	**3**
Chapter 3	The Role of Lanthanum as Promoter in Heterogeneous Catalysis	**5**
Chapter 4	Concluding Remarks	**25**
References		**27**
Index		**37**

PREFACE

The interest for studying lanthanum compounds has been growing in recent years, due to their attractive properties for industrial and technological purposes. They found several applications in different fields, such as component in various optical, electrical and magnetic applications, as well as catalytic materials. In this case, they are widely used as catalysts, supports and dopants. As a dopant, lanthanum is often used for stabilizing the gamma phase of alumina, in catalysts designed for high temperature applications, such as in automotive three-way catalytic process, in combustion for gas turbines and boilers and in hydrocarbon reforming. Besides alumina, other supports and catalysts have been improved by lanthanum, such as Ni/BaTiO$_3$, silica, ceria, titania, HCa$_2$Nb$_3$O$_{10}$, K$_{1-x}$La$_x$Ca$_{2-x}$Nb$_3$O$_{10}$ and iron oxides, for different reactions. The lanthanum addition to hematite, for instance, changes the specific surface area, the resistance against reduction and also the activity, selectivity and resistance against coke deposition of the catalysts during ethylbenzene dehydrogenation. However, these effects largely depend on the amount of the dopant, on the preparation method as well as on the kind of precipitant. Lanthanum also improves catalysts for applications at low temperature such as nickel and copper-based solids or titania-supported gold for carbon monoxide oxidation, zirconia-supported platinum and tungstophosphoric acid for n-pentane isomerization, lanthanum and nickel supported on sepiolite for styrene hydrogenation and La-MCM-41 catalysts for styrene oxidation. These examples, and many others, make lanthanum a very promising dopant for heterogeneous catalysts for different applications.

Chapter 1

INTRODUCTION

Lanthanum compounds have found several applications in different fields because of their attractive properties for industrial and technological purposes. They are widely used, for instance, as a component in various optical, electrical and magnetic applications such as optical wave guide, optical filters and capacitors [1, 2]. Also, lanthanum oxide is an important material used to produce ceramic superconductors, because of its magnetic properties [3] and is largely used as a strengthening agent in structural materials [2]. Moreover, this compound has several applications in catalysis as active phase, support and promoter [4-6].

In addition, lanthanum compounds have been extensively investigated in recent years, since their importance in fuel cells was recognized. These compounds find applications as cathode, electrolytes and ceramic interconnects in solid oxide fuel cells (SOFC) [7-9]. It is well-known, for instance, that solid solutions based on perovskite-type oxide $LaGaO_3$, especially when doped with strontium or magnesium (LSGM), exhibit ionic conductivity higher than the classical stabilized zirconia electrolyte [10-12]. Also, they have been recognized as potential candidates as electrolyte materials for SOFCs operating at intermediate temperatures [13, 14]. On the other hand, $LaCoO_3$ perovskite and lanthanum manganite ($LaMnO_3$)-based materials fit the requirements for SOFC cathodes, although some solid state reaction with yttria-stabilized zirconia occurs [9, 11, 15, 16]. Besides, lanthanum chromite-based perovskite oxides ($LaCrO_3$) have been pointed out as promising interconnect materials for SOFC. To fit the requirements for this application, they are normally doped with lower valence alkaline ions such as Ca^{2+}, Mg^{2+} and Sr^{2+} at the La^{3+} or Cr^{3+} sites [17, 18].

Chapter 2

APPLICATIONS OF LANTHANA AS CATALYSTS AND SUPPORTS

Because of their basic properties and their refractory nature, lanthanum compounds have found expanding applications in several reactions both as supports and as catalysts. These uses largely depend on the preparation methods and on the pretreatment conditions, which determine the basic character and then the catalytic properties [19, 20].

Some examples of lanthana as catalysts included the dehydration/ dehydrogenation of ethanol [21], the NOx decomposition [4], the NO reduction by methane [22], the oxidative coupling of methane [23, 24] and the combustion of diesel soot [25]. As perovskites, the lanthanum compounds showed catalytic activity in fuel combustion [26], carbon monoxide hydrogenation [27], dry reforming of methane [28-30] and partial reforming of methane [31].

As catalytic supports, lanthana has been used in the synthesis of methanol from syngas [32], in the ethane hydrogenolysis and cyclopropane hydrogenation [5], in the n-heptane dehydrocyclization [33] and in the oxidative coupling of methane [34], as well as in diesel soot elimination [35] and in dry reforming of methane [36].

Chapter 3

THE ROLE OF LANTHANUM AS PROMOTER IN HETEROGENEOUS CATALYSIS

Lanthanum is widely used as dopant in several fields, such as for improving the corrosion resistance of hot-dipped galvalume coating steel wire [37] or the electrical conductivity of cathode materials (LiNiVO$_4$ and LiCoVO$_4$) for secondary litium batteries [38, 39], among other applications.

As far as catalysis is concerned, lanthanum is largely used as a dopant both in the support and in the active phase, as shown in Table 1. One of the most common applications is by far in stabilizing the gamma phase of alumina, which is extensively used as catalysts and supports, due to its high specific surface area, surface acidity and defects in its crystalline structure [40-43]. For the automotive three-way catalytic (TWC) process, for instance, γ-Al$_2$O$_3$ provides a high and stable surface area for dispersion of the precious metals [44, 45]. However, during the normal running of an engine, the catalytic bed can rise to over 1000 °C and the gamma phase can go on phase transition to produce the thermodynamically most stable α-Al$_2$O$_3$ [41]. Also, for the combustion catalysts for gas turbines and/or boilers, working for several years at 1000-1400 °C and for methane steam reforming catalysts, operating at 1000 °C under high pressure, the thermal stability of the catalysts is much needed [46].

Therefore, numerous studies have been addressed to improve the thermal stability of alumina, by using new preparation methods or by adding foreign substances to the solids [41-43, 46-50]. These additives act kinetically delaying sintering and phase transition, since these processes occur thermodynamically. It is generally accepted that the phase transformation of transition aluminas occurs through a dehydroxylation process, during which the phase transition to α-Al_2O_3 proceeds by sintering of transition aluminas [46, 47]. The additive cations occupy both surface and bulk sites preventing atomic diffusion, which causes the phase transition [46]. Zr^{4+}, Ca^{2+}, Th^{4+} and La^{3+} species are able to inhibit these processes while other species, such as In^{3+}, Ga^{3+} and Mg^{2+}, accelerate the transformation [47, 48]. Among the stabilizers of transition alumina, lanthanum is the most studied one [41-43, 46-50]. It has been proposed [47, 49, 50] that this stabilization is related to the production of lanthanum aluminate on the surface which is able to retard the surface diffusion by the formation of a coherent interface, that causes a strong interaction responsible for the thermal stability. However, other workers [51] have proposed that this stabilization is related to lanthana surface layers, which decrease the surface energy, the driving force for sintering. In spite of the several works that have been carried out [47, 49-51], the chemical state of additives in alumina and their effects on sintering and on phase transitions remain controversial and ambiguous and the mechanism of stabilization has not been completely established.

The control of thermal stability enables alumina and other oxides to be used as support for applications at high temperatures such as in combustion, hydrocarbon reforming and catalytic ceramic membranes. Moreover, other advantages of using lanthanum as dopant have been recognized, besides thermal stability, which has been motivated its applications for other purposes, as shown in Table 1.

3.1. THE STABILIZATION OF COMBUSTION CATALYSTS WITH LANTHANUM

Catalytic combustion is more efficient to produce clean mixtures and less pollutant than conventional flame combustion. It can be used in gas turbines, boilers, aircrafts, afterburners, heat extraction, domestic heaters and others [52]. The process has been extensively studied since the 1970's, especially the use of oxidation catalysts in gas turbines, which operate in the range of 1000-1400 °C [52, 53]. In this case, the thermal stability is crucial. For this purpose, several materials have been investigated for methane oxidation in excess oxygen, such as Pd/Al_2O_3, Pt/Al_2O_3, Co_3O_4, $LaCoO_3$, $La_{0.5}Sr_{0.5}CoO_3$, $Cu/La-Al_2O_3$ and $Sr_{0.8}La_{0.2}MnAl_{11}O_{19}$ [54]. Among them, the precious metals showed the highest activities being palladium the most promising one, which was 30 fold more active than Co_3O_4, the most active oxide.

Table 1. Applications of lanthanum as dopant in heterogeneous catalysis

Catalyst	Reaction	Temperature range (°C)	Improvement	Ref.
Cu/La-Al$_2$O$_3$	combustion	1000-1400	thermal stability and metal activity	54, 55
Pd/La-Al$_2$O$_3$	combustion	1000-1400	thermal stability and metal sintering delay	56, 57
Ni/La-Al$_2$O$_3$	dry reforming	650-850	activity and selectivity	58, 59, 60
	partial oxidation of methane	700-800	support strength	61
	methane steam reforming	500-600	low ignition temperature and less coke	62
	propane steam reforming	<1000	prevent nickel agglomeration	63, 64
	diesel oxidative reforming	400-550	activity and selectivity	
Pt/La-Al$_2$O$_3$	diesel oxidative reforming	650-800	thermal stability and selectivity	65, 66
Ni/La-BaTiO$_3$	dry reforming	650-800	thermal stability and selectivity	65, 66
Co(O)/La-SiO$_2$	dry reforming	700-800	specific surface area, activity	68
Ni-La/Al$_2$O$_3$-SiO$_2$	ethanol steam reforming	600	activity and less cobalt sintering	69
Pt, Rh/CeO$_2$/La$_2$O$_3$/Al$_2$O$_3$	reactions in automobile exhaust	400	selectivity, stability, less coke	70
Pd/Al$_2$O$_3$-La$_2$O$_3$	reactions in automobile exhaust	500	thermal stability and oxygen storage capacity	71
Ce$_{1-x}$La$_x$O$_{2-x/2}$	reactions in automobile exhaust	500	thermal stability and oxygen storage capacity	74, 75
	reactions in automobile exhaust	---	thermal stability	73

Table 1. (Continued).

Catalyst	Reaction	Temperature range (°C)	Improvement	Ref.
Pd/γ-Al$_2$O$_3$-La$_2$O$_3$	hydrocarbons dehydrogenation (membranes)	1000	thermal stability	76
La-TiO$_2$	pCBA photodecomposition	--	photoactivity	82
	phenol photodegradation	--	photoactivity	83
	methyorange photodegradation	--	photoactivity	79
	acetone oxidation	--	photoactivity	84
La-doped HCa$_2$Nb$_3$O$_{10}$	water splitting	--	photoactivity	85
CuO-NiO-La	CO oxidation	250	activity	103
Au/TiO$_2$-La	CO oxidation	--	activity	
Pt-La-TPA/ZrO$_2$	n-pentane isomerization	200	activity	105
Ni-La/sepiolite	styrene hydrogenation	90	activity	106
α-Fe$_2$O$_3$-La	ethylbenzene dehydrogenation	530	activity, selectivity	6, 90, 91
La-MCM-41	styrene oxidation	60	activity	106

The Cu/La-Al$_2$O$_3$ catalyst also demonstrated to be promising and has been subject of several studies because of the La-stabilized alumina, which showed a high surface area [54]. In addition, Jiang et al. [55] have shown that the addition of lanthanum increases the catalytic activity and the copper dispersion on the support; besides, this dopant makes the copper reduction easier and improves the surface oxygen desorption. However, the Pd/Al$_2$O$_3$ catalyst has been recognized as a superior catalyst for years [56, 57]. Chou et al. [57] have pointed out that the addition of lanthana to Pd/Al$_2$O$_3$ catalyst inhibited the activity decay at high temperatures; they concluded that this dopant not only increased the thermal stability of γ-Al$_2$O$_3$ but also delayed the palladium sintering upon high combustion temperatures, by inhibiting the oxygen desorption from PdO, due to an increase in the bond strength.

3.2. THE ROLE OF LANTHANUM IN IMPROVING THE CATALYSTS FOR HYDROCARBONS REFORMING

Considerable attention has been given to the reforming of hydrocarbons in order to find efficient and cost-effective technologies for hydrogen production for several applications, including fuel cells. The reforming processes are often performed at high temperatures which favor several kinds of catalyst deactivation such as sintering, phase changes, coke deposition and others. Therefore, most of the studies are addressed to increase the thermal stability of the catalysts [58-61]. However, other advantages were found in adding lanthana and other oxides to the reforming catalysts, such as the increase of the activity and selectivity as well as the resistance against coke, as a consequence of changes in metal reducibility and dispersion as well as in support acidity [62].

Cheng et. al. [58] noted that the addition of lanthanum oxide increases the activity and the selectivity to syngas of Ni/Al$_2$O$_3$ catalysts in dry reforming of methane, performed in the range of 650-850 °C. This effect depends not only on the preparation method (co- or successive impregnation) but also on the order of adding the promoter. The impregnation of lanthana after NiO in alumina led to the best promotion action. This effect was related to the interaction between lanthana and nickel, modifying the methane and carbon dioxide adsorption as well as their dissociation in the active surface. Besides the improvement of the activity and selectivity, Slagtern et. al [59, 60] observed that lanthanum increases the support strength. They concluded that nickel enters into positions in the Al-O spinel blocks of the subsurface

$LaAl_{12}O_{19}$ phase, decreasing the reducibility and thus the activity increases for the reforming reaction as compared to nickel on unmodified alumina. The modified catalysts were also more stable than the unmodified catalyst but the stability was sensitive to the pretreatment conditions of the catalysts.

The beneficial effect of lanthana to Ni/Al_2O_3 catalysts was also noted in partial oxidation of methane. Lanthana was able to low the ignition temperature of the reaction and to inhibit the coke deposition on the catalyst surface in combination with CaO [61]. Moreover, the addition of lanthana decreases the interaction of NiO and Al_2O_3 to produce new species of $LaNiO_3$ after calcination at 800 °C. However, the excess of lanthana leads to the production of free NiO phase which decreases the activity and increases coke formation.

For methane steam reforming, it was noted [62] a large effect of small additions of several metal oxides on the hydrogen reduction of precipitated NiO on alumina. Thus, the oxides with more difficulty to be reduced, such as ZrO_2 and La_2O_3, make slow the nickel oxide reduction. This was related with the ability of the dopant to retain water on the surface in enough amounts to prevent chemical reduction and agglomeration even at high temperatures, typical of commercial steam reforming operations.

Besides methane, other hydrocarbons have been considered as suitable feedstocks for hydrogen production. In this case, the catalyst deactivation is more severe, as compared to natural gas steam reforming processes. The low price of nickel-based catalysts, commercially used for steam reforming of natural gas, has motivated several studies devoted to optimize these catalysts for steam reforming of heavier hydrocarbons than methane, including gasoline and diesel [63]. Natesakhawat et al. [63, 64] pointed out that small amounts (2 wt.%) of lanthanide elements added to alumina-supported nickel largely improve the catalytic activity and stability in steam reforming of propane, due to an increase in catalyst reducibility, nickel surface area and resistance to deactivation.

Alvarez-Galvan et al. [65, 66] studied platinum and nickel phases deposited on alumina modified with lanthanum and cerium oxides for the oxidative reforming of diesel, using hexadecane, decalin and tetralin, selected as models hydrocarbons for the paraffin, cycloparaffin and aromatic fractions present in diesel. This feedstock shows several advantages such as high volumetric and gravimetric hydrogen density, low cost, availability and ease handling with an infrastructure already in place. However, diesel is a complex mixture of hydrocarbons with a boiling range between 200 and 380 °C, requiring very high temperatures for the steam reforming. In addition, catalysts must be active for hydrocarbons of different properties and resistant to sulfur and coke deposition [67]. It was found that the activity depends on the kind of metal, being nickel more active than platinum and also more selective to hydrogen. Nickel phase was preferentially deposited on ceria particles while platinum was deposited mostly on alumina, because of the isoelectric point of the oxides. The introduction of lanthanum increased the specific surface area by 20 % after thermal stabilization and decreased ceria dispersion. Also, lanthana inhibited the formation of the γ-Al_2O_3 phase. The catalysts in which platinum crystallites were deposited on the alumina substrate, covered by a lanthana monolayer, gave rise to an increase in stability toward hydrogen production.

Other supports can be improved by lanthana, as was also observed by Xiancai et. al. [68] for dry reforming catalysts. They have found that the introduction of lanthanum to $Ni/BaTiO_3$ catalyst increased the specific surface area and the metal dispersion up to an optimum amount (1.5 % wt); the excess of lanthanum caused the pore blockage decreasing the specific surface area. The catalytic activity in dry reforming was also improved due to lanthanum, a fact that was assigned to an increase of oxygen vacancies, which help the reaction; however, the catalytic activity had a decreasing tend with the additional increment of lanthanum. Moreover, lanthanum favored nickel reduction.

A positive effect of lanthana was also observed by Bouarab et al [69] for silica-supported cobalt in methane dry reforming. They found a direct relationship between basicity and activity and concluded that lanthana adjusts the acid function of the unpromoted catalyst, which is responsible for coke production. Also, lanthana prevents cobalt phase sintering, by avoiding particles coalescence.

The advantage of adding small amounts of lanthanum to silica-based supports was also noted by Zhang et al. [70], studying catalysts for ethanol steam reforming. They prepared nickel-based catalysts supported on $Al_2O_3.SiO_2$ modified with lanthanum, cobalt, copper, zinc or yttrium. The addition of lanthanum resulted in catalysts with higher selectivity to hydrogen and lower selectivity to carbon monoxide as compared to cobalt-containing solids. Low amounts of lanthanum (5 % wt) inhibited the crystal growth of nickel and favored the reduction of nickel oxide but higher amounts (15 % wt) make the nickel reduction more difficult. The optimum catalyst ($30Ni2La/Al_2O_3.SiO_2$) showed long-term stability (100 h), providing high hydrogen selectivity and low carbon monoxide, besides high resistance against coke production at low temperature.

3.3. THE DOPING OF THE THREE-WAY CATALYSTS WITH LANTHANUM

Catalytic automotive pollution control has been subject of numerous researches in recent years. The abatement of pollutants in automobile exhausts is carried out by the three-way catalysts (TWC) that are constituted of ceria-based compounds in cooperation with precious metals dispersed on alumina [71, 72]. Ceria has the important role of acting as a storage oxygen component because of its ability to store and supply oxygen to the noble metals on which the oxidation of carbon monoxide proceeds. This property depends on the mobility of oxygen in ceria lattice and can be affected by the presence of dopants in the lattice [73]. It is known that the addition of trivalent species such as lanthanum, for instance, improves the oxygen mobility due to the anionic vacancies that were formed to keep the lattice charge neutrality [71, 73, 74]. Furthermore, lanthanum improves the thermal stability of γ-Al_2O_3, avoiding its transformation to α-Al_2O_3 [41, 74, 75]. By studying ceria-lanthana-based TWC promoters by EXAFS spectroscopy, Deganello et al. [73] detected the presence of a fluorite-like $Ce_{1-x}La_xO_{2-x/2}$ solid solution (that is important for oxygen storage capacity) and of the perovskite-like $LaAlO_3$ compound, suggesting a competition between the two different environments in hosting lanthanum. Also, lanthanum was found in alumina phase, supporting the hypothesis that it plays an important role in the thermal stabilization of gamma-alumina.

3.4. THE ACTION OF LANTHANUM ON THE THERMAL STABILITY OF CATALYTIC MEMBRANES

Ceramic membranes play an important role in the hydrogen technology, in both separation and filtration, besides catalytic reaction. As compared to polymeric membranes, they show several advantages such as chemical stability, long life and good defouling properties [45, 76].

One of the most important catalytic applications of ceramic membranes is in dehydrogenation of hydrocarbons, using Pd/γ-Al_2O_3, which operates at high temperatures (around 1000 °C). This requires the stability of this phase, avoiding sintering and its transformation to α-Al_2O_3. In fact, above 800 °C, the BET surface area of pure alumina membranes begins to decrease and above 1000 °C the pore structure, suitable for membrane applications, is destroyed limiting high temperature applications [76]. In order to overcome these drawbacks, Chen et al. [76] studied the effect of lanthanum on the thermal stability of alumina-supported palladium membranes and found that this dopant improves the average pore diameter, the pore volume and the BET area of the ceramic membranes. The higher thermal stability of the lanthanum-doped membrane was related to the retarded phase transition from γ-phase to α-phase of alumina.

3.5. THE ADDITION OF LANTHANUM TO PHOCATALYSTS

The degradation of organic compounds can be efficiently catalyzed by several semiconductors with suitable band-gap. Among them, titanium oxide has been pointed out as the most promising catalyst for detoxification of low concentration organic pollutants [77]. It is well-known [78, 79] that doping the semiconductors with various transition metals ions may lead to an increase of efficiency of the photocatalytic systems. It is also known [80] that the reactivity is a complex function of the dopant concentration, the energy level of the dopants in the lattice, their distribution and electronic configuration, the electron-donor concentration and the light intensity. Due to the ability of lanthanide ions to form complexes with several Lewis bases (acids, amines, aldehydes, alcohols, thiols and others) by the interaction of their functional groups with the f orbitals of the lanthanides [78], it is expected that the addition of lanthanide ions to titania causes an increase of concentration of organic pollutant on the semiconductor surface [81]. However, the La^{3+} ions act as p-type dopant (acceptor centers) which trap photoelectrons and attract holes, producing recombination centers. Because of the smaller ionic radius of Ti^{4+}, these ions are supposed to enter into lanthana lattice and more hydroxide ions would be adsorbed in order to compensate the charge. The hydroxide ions can accept holes generated by UV irradiation to form hydroxyl radicals, which oxidize further adsorbed molecules. Therefore, the photoinduced charge carriers recombination can be suppressed. During heat treatment of the catalyst, the charge imbalance caused by titanium in lanthana lattice may cause a partial reduction of Ti^{4+} to Ti^{3+}, the latter acting as a charge-carrier recombination center and advances the recombination of electron–hole pairs [82].

Kim at al. [82] have noted a proportional dependence between the photoluminescence intensity and the photocatalytic activity of lanthanum and titanium mixed oxides. The sample with 30 % wt of La exhibited the highest activity for pCBA (4-chlorobenzonic acid) photodecomposition. The role of lanthanum in increasing the photoactivity was related to the changes on titania surface, caused by lanthanum, which retard the recombination of photoexcited electron/hole pairs, resulting in a higher photoactivity in the stronger photoluminescence intensity. Liqiang at al. [83] also have found a relationship between the photoluminescence spectra and the photocatalytic activity for phenol degradation namely, the stronger vacancies and defects, the higher the photocatalytic activity. They also noted great inhibition on titania phase transformation by lanthanum.

The effect of lanthanum on the properties of titania nanopowders catalysts was studied by Xu et al. [79] using the photocatalytic degradation of methyl orange. They have found that lanthanum doping strongly increased the photoactivity of titania, being 0.5 at% the optimal concentration. Lanthanum also decreased the titania particle size and the distribution of particle sizes became more uniform. The doped powders were made off anatase and rutile phases and the amount of anatase phase decreased and then increased with an increase of lanthanum amount.

Lin and Yu [84] showed the effect of rare earth on the photoactivity of titania in the oxidation of acetone. They found a positive effect for lanthana (0.5 wt.%) or yttria (0.5 wt.%) but a negative effect for ceria. These differences were related to the change in the amount of surface hydroxyl groups resulting from the interaction between the rare earth oxides and titania. The presence of these rare earth oxides can inhibit anatase to rutile transformation at elevated temperatures.

The beneficial doping of lanthana was also observed for photocatalytic splitting of water into hydrogen and oxygen over $HCa_2Nb_3O_{10}$ catalyst [85]. This process has received attention in recent years because of its potential application for the conversion of solar energy into clean-energy hydrogen fuel [86]. Most of the active photocatalysts for water splitting have the perovskite-related structure ABO_3, whose activity depends on the electronic band structure and on the bulk crystal structure. It is well-known [85] that the constituent elements affect the shape of the valence and conduction band of a photocatalyst, by affecting the electron–hole separation process after absorption of light with wavelengths shorter than its band gap energy. The activity of these catalysts can be highly improved by partial substitution on A and/or B sites, with only small changes in the structure. Moreover, they contain semiconducting host layers and interlayer alkali cations. Charge separation takes place in the host layers upon ultraviolet irradiation. The generated electrons and holes show high reductive and oxidative reactivity, generating photocatalytic properties. The interlayer guests are ion-exchangeable with various foreign species [85]. It is known [85, 87, 88] that the addition of small amounts of lanthana increases the activity of these catalysts. Huang et al. [85] studied the properties of $K_{1-x}La_xCa_{2-x}Nb_3O_{10}$ (x= 0, 0.25, 0.5, and 0.75) and their derivates $H_{1-x}La_xCa_{2-x}Nb_3O_{10}$ with layered perovskite structure by substituting Ca^{2+} with La^{3+} in $Ca_2Nb_3O^{2-}_{10}$ in water splitting. Platinum metal nanoparticles (active sites) were introduced directly into the interlayer of compound. They concluded that the band gap energy of these compounds was changed by lanthanum and by doping with suitable amount of lanthanum the activity was increased. The photocatalyst $H_{0.5}La_{0.5}Ca_{1.5}Nb_3O_{10}$ showed the highest activity, which was improved even more by adding platinum to the solid.

3.6. IMPROVEMENTS OF CATALYSTS FOR ETHYLBENZENE DEHYDROGENATION BY LANTHANUM

Lanthanum has been widely investigated [6, 89-91] as promoter for hematite-based catalysts for ethylbenzene dehydrogenation in the presence of steam. This process is the main commercial route to produce styrene, a high-value chemical, largely used for the synthesis of many polymers such as polystyrene, styrene-acrylonitrile, acrylonitrile-butadiene-styrene and styrene-butadiene latex. The reaction is endothermic, reversible and equilibrium limited, being favored by low partial pressure of ethylbenzene and hydrogen as well as high temperatures. The commercial operations are normally carried out under high amounts of steam, which promote the forward reaction, supply heat to reaction and limit the deposition of coke by gasification, besides acting as oxidant agent, producing an appropriate oxidation state in iron oxide [92, 93].

The catalysts often used in commercial processes consist in hematite (α-Fe_2O_3) and promoters such as potassium oxide and chromium oxide, among others [92]. It is well-known [92-97] that potassium acts as electronic promoter increasing the iron activity and also contributes to decrease the coke formation. On the other hand, chromium acts as textural promoter increasing the specific surface area of hematite [92, 93]. These catalysts are cheap and very active and selective, but have some drawbacks such as the low specific surface area and the instability of the active oxidation state of iron; hematite (α-Fe_2O_3) is preferred for styrene production, but it tends to transform to compounds with lower oxidation states and even to elemental iron, both of which catalyze carbon formation and dealkylation [92, 97-99] Moreover, they deactivate with time and is susceptible to poisoning by halides and residual organic chloride impurities [93]. However, the most serious deactivation is caused by the lost of potassium, which can migrate in two directions during commercial operations; potassium chloride can be found both in the water layer of the condensed product and in the catalyst pellets. The major migration of potassium occurs within the catalyst pellets, due to the reaction endothermicity which makes the centre of the pellets colder than the periphery. Other disadvantages of the commercial catalysts include the large amounts of steam used in industrial units increasing the operational costs and the toxicity of the chromium compounds causing damage to the humans and to the environment [93, 97-99].

With the aim of obtaining less toxic and potassium-free catalysts to ethylbenzene dehydrogenation, several works have been carried out [6, 90, 91, 93, 100-102]. An efficient and promising strategy to get more active and selective iron-based catalysts is to add different dopants which can replace potassium and/or chromium.

The effect of lanthanum on the properties of hematite-based catalysts for ethylbenzene dehydrogenation was studied by Santos et. al. [90]. They used three preparation methods by changing the way of mixing the reactants while the lanthanum to iron molar ratio was kept equal to 0.1. In the first method, the iron and lanthanum nitrate solutions and the ammonium hydroxide solution were added to a beaker with water, at room temperature. In the second one, both iron nitrate and lanthanum nitrate solutions were added to a beaker containing the ammonium hydroxide solution. The third method consisted in the inverse procedure. A solid prepared by the mechanical mixture of the two metal oxides and a lanthanum-free sample were also prepared. It was found that the preparation method strongly affects the properties of the catalysts. The samples showed different particle sizes and specific surface areas as well as different resistance against reduction. The most active catalyst in ethylbenzene dehydrogenation was obtained by adding the iron and lanthanum nitrate solutions to an ammonium hydroxide solution. This solid was also able to produce low amounts of coke and has the advantage of being non-toxic.

In order to optimize these catalysts, the effect of lanthanum amount on the properties of hematite was also studied [6]. The samples were obtained by the sol-gel method, from lanthanum nitrate, iron nitrate and ammonium hydroxide using lanthanum to iron molar ratios of 0.2 and 0.1. The addition of lanthanum increased the specific surface area of hematite (17 $m^2.cm^{-1}$) and the highest value was obtained for the sample with La/Fe= 0.1 (115

$m^2 \cdot g^{-1}$). This was related to the action of this dopant as a spacer on the solid surface keeping the particles apart. Due to the large radius of La^{3+} ions (1.05 A), in comparison with Fe^{3+} ions (0.64 A), they cannot be enter into the iron oxide lattice and thus must produce a segregated phase. During the ethylbenzene dehydrogenation, however, the specific surface area decreased, due to particles growing. It was noted, by XPS, that the surface was rich in lanthanum both before (La/Fe= 2.839) and after (La/Fe= 3.137) reaction. Therefore, one can suppose that the catalysts are made off hematite particles with small lanthana particles on the surface, as shown in Figure 1.

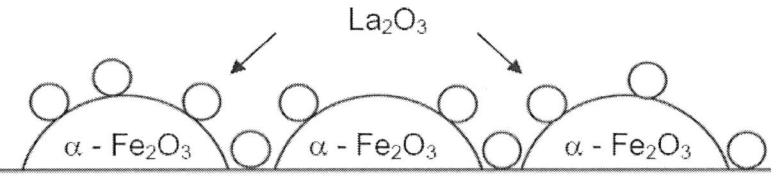

Figure 1. Scheme showing the anti-sintering action of lanthana in iron oxide.

The catalyst with La/Fe= 0.1 showed the best performance and this was related to its high resistance against reduction, stabilizing the Fe^{3+} species (active phase) as well to its ability to prevent coke deposition. This effect can be compared to the action of other dopants found by other authors [93, 97]. These promoters were supposed to increase the activity of iron oxide by electron transfer in solid-gas interface making the iron-oxygen system more polarizable, increasing the basicity of the Fe^{3+} active sites. These catalysts showed high selectivities to styrene, being the sample with the lowest amount of lanthanum (La/Fe= 0.1) the most selective. For all cases, toluene was produced in higher amount than benzene [6].

The catalyst with La/Fe= 0.1 was also prepared using sodium hydroxide and sodium carbonate, by the same experimental method. In these cases, big crystals were produced. The effect of the precipitant agent on the particle size was studied by Mössbauer spectroscopy. For all fresh catalyst a doublet was observed (Figure 2) probably related to superparamagnetic hematite (α-Fe_2O_3). The solid prepared with ammonium hydroxide showed two magnetic components in the spectra, related to the "core" and "shell" of the particles. The catalyst obtained with sodium carbonate showed the biggest particle size while that precipitated with ammonium hydroxide showed the smallest one. Only magnetite was detected for all spent catalysts. The solid precipitated with sodium hydroxide showed the highest amount of Fe^{+2}, indicating that they are more reducible than that prepared with ammonium hydroxide. As expected, the highest specific surface area was shown by the solid prepared with ammonium hydroxide, indicating that sodium ions favored the particle growing and then the specific surface area decreasing. After reaction, the specific surface area values remaining low, except for the sample without sodium, which went on sintering during ethylbenzene dehydrogenation [6].

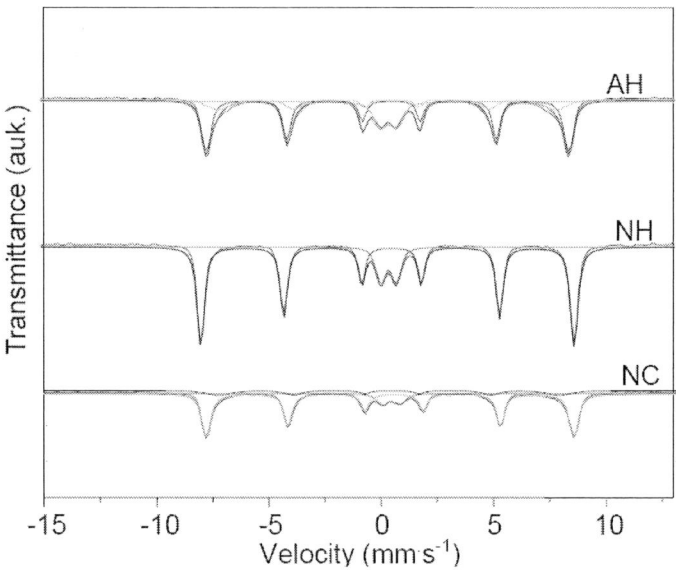

Figure 2. Mössbauer spectra of lanthanum-doped hematite prepared with ammonium hydroxide (AH), sodium hydroxide (NH) and sodium carbonate (NC).

It also noted that sodium ions made the reduction of Fe^{3+} species more difficult, independently of the anion (carbonate or hydroxide) of the precipitant. On the other hand, the solid prepared with sodium hydroxide showed a lower resistance against reduction than that obtained with sodium carbonate. Sodium ions favored the production of less active catalytic sites, a fact that was related to the easiness of the sodium-containing solids to produce Fe^{2+} species, during reaction. Moreover, sodium carbonate increased the selectivity towards styrene while sodium hydroxide caused a decrease [6].

No lanthanum was detected on the surface of the samples prepared with sodium carbonate, which explains its low specific surface area; the surface was rich in sodium, that partially covered the iron active sites, explaining the low intrinsic activity of the catalysts. On the other hand, the surface of the solid prepared with ammonium hydroxide showed the highest amount of

The Role of Lanthanum as Promoter in Heterogeneous Catalysis   19

lanthanum, as compared to the other samples; this explains the high specific surface area of the fresh catalyst. During ethylbenzene dehydrogenation, the particles of lanthanum oxide grew and lost their anti-sintering action, causing a decrease of the specific surface area of the spent catalyst. The presence of lanthanum on the surface also caused an increase of activity per area, which showed low values in the solids poor in lanthanum on the surface. From these results, it can be concluded that sodium has a negative effect on the specific surface area and on the catalytic activity of lanthanum-doped hematite [6].

Among the precipitants used for preparing lanthanum-doped hematite, ammonium hydroxide produced the best catalyst. However, its use in industry is not allowed due to the environment and human health restrictions. Therefore, new catalysts were prepared to study the effect of the order of mixing the reactants on the properties of hematite with lanthanum, using potassium carbonate as precipitant [91]. Samples were prepared by the sol-gel method by hydrolysis of iron and lanthanum nitrate with a potassium carbonate solution to get solids with lanthanum to iron molar ratio of 0.1. The first sample was obtained by the addition of the reactants solutions on a beaker with water while the others were prepared by changing the order of mixing the reactants: by adding the precipitant on the solution of iron and lanthanum nitrate and by the inverse procedure. The sample prepared with ammonium hydroxide showed only hematite while those obtained with potassium carbonate showed several phases, depending on the preparation method; these solids showed lower specific surface areas than that prepared with ammonium hydroxide; the solid with the lowest specific surface area was obtained by the addition of the reactants to water but no significant difference was found between the samples prepared by the other methods. During ethylbenzene dehydrogenation, the specific surface values increased, indicating that the phase changes led to the formation of pores and/or particles of smaller sizes.

The resistance against reduction was also changed by the order of mixing the reactants; the sample prepared by adding the precipitant to the metallic precursors was the most susceptible to reduction while the sample prepared by adding the reactants on water showed the highest resistance against reduction, showing that the Fe^{3+} species are stabilized in this solid, as inferred by Mössbauer spectroscopy, as shown in Figure 3. In fact, the spectrum of the potassium-containing sample showed two sextuplets, with hyperfine parameters typical of α-Fe_2O_3 and two different particle sizes. A central doublet was also noted, associated to superparamagnetic α-Fe_2O_3 and/or to paramagnetic ferric ions in La_2O_3 lattice [91].

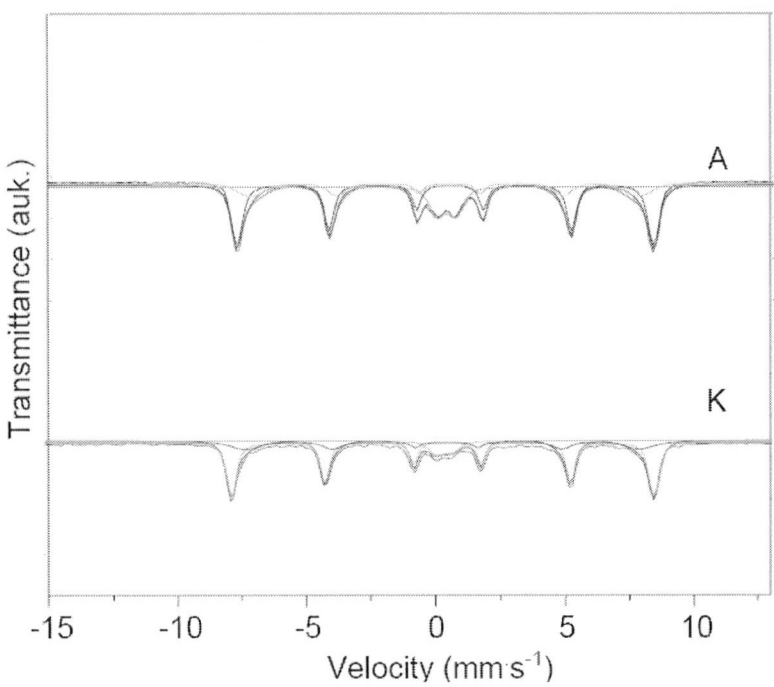

Figure 3. Mössbauer spectra of lanthanum-doped hematite prepared with ammonium hydroxide (AH) and potassium carbonate (KC).

The order of mixing the reactants and the presence of potassium affected the activity of the catalysts, potassium increased the activity by increasing the intrinsic activity (a/Sg). Also, the styrene selectivity achieved 100 % due to potassium and no benzene and toluene were produced over the potassium-containing catalysts. These solids produced higher amounts of coke than the potassium-free sample; therefore, lanthanum is much more able to prevent coke in the absence of potassium and is more efficient than potassium. However, this coke seems not to be harmful to the catalysts, since it is not related with the conversion drop. The catalyst prepared by adding the reactants on water was the most active one, probably due to its highest intrinsic activity and resistance against reduction, due to the stabilization of Fe^{3+} species, which are responsible for the activity of the catalysts in the reaction. This sample has lanthanum (La/Fe= 0.884) and a high concentration of potassium (K/Fe= 12.229) on the surface, suggesting the production of potassium oxide, which is believed to increase the activity of iron in ethylbenzene dehydrogenation [92-97]. This catalyst was more active and selective than a commercial one which makes it a candidate for industrial applications.

From these results, one can see that when lanthanum is present together with alkaline ions such as sodium or potassium, its interaction with iron is increased, as shown in Table 2, obtained from TPR results [6, 90, 91]. When ammonium hydroxide was used (AH sample) peaks at 390 and 680°C, assigned to hematite reduction to produce magnetite and to magnetite to form metallic iron, respectively, were found. Besides, the peak at 500°C is related to lanthana reduction. By comparing these values with the other samples, we can conclude that sodium and potassium ions lead to the stabilization of Fe^{+3} species making them more resistant against reduction. The curves of the samples prepared with sodium-based precipitants showed a single peak at high temperatures, related to the simultaneous reduction of iron and lanthanum species. In the case of the solids obtained with sodium carbonate or potassium carbonate, the peaks were shifted to even high temperatures, indicating that the carbonate ion contributes to produce solids even more resistant against reduction. As a whole, potassium caused a stronger interaction between lanthanum and iron than sodium.

Table 2. Reduction temperatures obtained from TPR curves from ref [6, 90, 91] of lanthanum-doped hematite prepared with ammonium hydroxide (AH), sodium hydroxide (NH), sodium carbonate (NC) and potassium carbonate (KC)

Samples	1st peak (°C)	2nd peak (°C)	3rd peak
AH	390	500	680
NH	590	710 (shoulder)	---
NC	610	--	---
KC	680	890	---

Lanthanum can be found on the surface or in the bulk of hematite (as a segregated phase), depending on the precipitant. It was predominantly on the surface when ammonium hydroxide was used. The presence of sodium or potassium caused its decrease on the surface and this effect was stronger in the sodium-containing samples. This effect was even more stronger for the sample prepared with sodium carbonate, in which case the solid showed only sodium and iron on the surface (Na/Fe = 26,155).

A comparative study of the role of zirconium, aluminum, lanthanum and neodymium as dopants for dehydrogenation catalysts was carried out by Ramos at al. [89]. Regarding activity, they found that lanthanum was the most efficient dopant, making hematite three times more active than the pure oxide. However, lanthanum decreased the selectivity from 92 % (pure oxide) to 88 % (doped hematite).

3.7. OTHER USES OF LANTHANUM AS DOPANT IN HETEROGENEOUS CATALYSIS

Lanthanum is widely used as dopants in several other catalysts for different applications at low temperatures. In these cases, other advantages than thermal stability were found such as an increase of activity and selectivity.

El-Shobaky et al. [103] studied the effect of lanthanum amount (2, 4 and 6 mol%) on the physicochemical, surface and catalytic properties of nanocrystalline CuO-NiO solids prepared by sol-gel method, in order to find efficient catalysts to carbon monoxide oxidation with oxygen, for exhaust emission control, carried out up to 250 °C. They have found that by adding 6 mol% of lanthanum, the CuO crystalline disappeared and the degree of crystallinity of NiO phase decreased. Also, doping caused a decrease of surface concentration of both copper and nickel species as well as of surface excess oxygen. By doping 0.25 % La, followed by calcination at 400 and 600 °C, a progressive increase in the catalytic activity of the doped solid catalysts was noted reaching to a maximum limit at 4 mol%. Some lanthana added to the solids calcined at 400 and 600 °C, was dissolved in NiO and CuO lattices and other portion interacted with copper oxide yielding lanthanum copprate (La_2CuO_4) and lanthanum copprite ($LaCuO_3$) compounds. The doping did not cause any change in the activation energy of the catalyzed reaction.

Yu et al. [104] also have found a positive effect of lanthanum for carbon monoxide oxidation catalysts (Au/TiO_2). They noted that lanthanum improved the activity, a fact that was assigned to the presence of lanthanum promoting the reactivity of CO adsorbed on gold sites and to the formation of a second active phase on the surface. It was observed that lanthanum increased the specific surface area and restrained the growth of titania particles.

Another application of lanthanum as dopant for low temperature catalysts was demonstrated by Xu et. al. [105] in n-pentane isomerization, carried out at 200 °C. They had proposed an alternative catalyst based on zirconia-supported platinum and tungstophosphoric acid (TPA), whose activity was found to be increased with the addition of lanthanum. Also, the optimal lanthanum amount strongly depended on the platinum loadings.

The promoting effect of several amounts of lanthanum on nickel supported on sepiolite for styrene hydrogenation was observed by Damyanova et. al. [106]. They found that lanthana was deposited on nickel particles, leading to a decrease in the probability to have adjacent nickel atoms on the surface. The improvement on the activity was related not only to the coverage of a larger fraction of the silicate substrate by lanthanum but also to the isolation of nickel atoms on the surface of nickel crystallites by lanthana patches deposited on it.

The addition of lanthanum also increased the catalytic activity of MCM-41 in the oxidation of styrene, as noted by Wangcheng at al. [107]. The La-MCM-41 catalysts exhibited higher activity than Ce-MCM-41, which increased with lanthanum content.

Chapter 4

Concluding Remarks

Lanthanum compounds have found various applications in different fields as important materials for industrial and technological purposes as well as for improving the properties of other materials. Regarding catalysis, lanthanum compounds are widely used as supports, catalysts and dopants in several reactions, both in laboratory and in industry. The beneficial effects of adding lanthanum to several materials is mainly related to the increase of the thermal stability, but several examples in literature have shown its ability in improving other properties of the catalysts such as activity, selectivity and resistance against reduction or coke deposition, as a consequence of changes in metal reducibility and dispersion as well as in support acidity. Lanthanum compounds also have been successfully used as different components of fuel cells, especially solid oxide fuel cells. Due to their suitable properties and versatility, the use of lanthanum in catalysis is expected to increase in next years, mainly for applications in fuel cells.

REFERENCES

[1] Q. Z. Wu, Y. Shen, J. F. Liao, Y. G Li. Synthesis and characterization of three-dimensionally ordered macroporous rare earth oxides. *Mater. Lett.* 58 (2004) 2688– 2691.
[2] S. Wang, W.Wang, Y. Qiana. Preparation of La_2O_3 thin films by pulse ultrasonic spray pyrolysis method. *Thin Solid Films.* 372 (2000) 50-53.
[3] R. Lailo, E. Lähderanta, L Säisä, Gy. Kovács, G. Zsolt. Optically induced changes in the magnetic properties of the ceramic superconductor $La_{1.8}Ba_{0.2}CuO_4$. *Phys. Rev. B: Condens. Mater. Phys.* 42 (1990) 347-353.
[4] X. Zhang, A. B. Walters, M. A. Vannice. NOx decomposition and reduction by methane over La_2O_3. *Appl. Catal. B: Environm.* 4 (1994) 237-256.
[5] G. R. Gallaher, J. G. Goodwin Jr., L. Guczi. Preparation and pretreatment effects on metal decoration in Rh/La_2O_3. *Appl. Catal.* 73 (1991) 1-15.
[6] M. S. Santos, S. G. Marchetti, A. Albornoz, M. C. Rangel. Effect of lanthanum addition on the properties of potassium-free catalysts for ethylbenzene dehydrogenation. *Catal. Today* 133-35 (2008)160-167.
[7] C. Xia, Electrolytes in : J.W. Fergus, R. Hui, X. Li, D. P. Wilkinson, J. Zhang (eds.) Solid Oxide Fuel Cells. Materials Properties and Performance. *CRC Press. Boca Raton.* 2009. pp. 59-60.
[8] T. Ishihara, Oxide Ion Conductivity in Perovskite Oxide for SOFC Electrolyte in: T. Ishihara (ed.) *Perovskite Oxide for Solid Oxide Fuel Cells.* Springer. London. 2009. pp.65-66.

[9] H. Yokokawa, T. Horita, Cathodes in: S; C.Singhal, K. Kendall (eds), High temperature solid oxide fuel cell: fundamentals, design and applications. *Elsevier. Oxford.* 2003. pp. 119-120.
[10] T. Ishihara, H. Matsuda, Y. Takita. Doped LaGaO$_3$ Perovskite Type Oxide as a New Oxide Ionic Conductor. *J. Am. Chem. Soc.* 116(9) (1994) 3801-3803.
[11] Z. Bi, Y. Dong, M. Cheng, B. Yi, Behavior of lanthanum-doped ceria and Sr-, Mg-doped LaGaO$_3$ electrolytes in an anode-supported solid oxide fuel cell with a La$_{0.6}$Sr$_{0.4}$CoO$_3$ cathode. *J. Power Sources* 161 (2006) 34-39.
[12] T. Ishihara, N. M. Sammes, O. Yamamoto, Electrolytes, in: S; C.Singhal, K. Kendall (eds). High temperature solid oxide fuel cell: fundamentals, design and applications. *Elsevier. Oxford.* 2003. Pp. 96-97.
[13] T. Ishihara, H. Matsuda, M. Azmi, Y. Takita. Effects of rare earth cations doped for La site on *the oxide ionic conductivity of LaGaO$_3$-based perovskite type oxide.* Solid State Ionics 79 (1995) 147-151.
[14] K. Q. Huang, R. S. Tichy, J.B. Goodenough. Superior Perovskite Oxide-Ion Conductor; Strontium- and Magnesium-Doped LaGaO$_3$: III. Performance Tests of Single Ceramic Fuel Cells. *J. Am. Ceram. Soc.* 81 (1998) 2581-2585.
[15] S. K. Lau and S. C. Singhal. Potential electrode/electrolyte interactions in solid oxide fuel cells. *Corrosion.* 85 (1985) 1-9.
[16] F. Bidrawn, G. Kim, N. Aramrueang, J. M. Vohs, R. J. Gorte. Dopants to enhance SOFC cathodes based on Sr-doped LaFeO$_3$ and LaMnO$_3$. *J. Power Source.* 195 (2010) 720-728.
[17] T. Horita, LaCrO$_3$-based perovskite for SOFC interconnects in: T. Ishihara (ed.) Perovskite Oxide for Solid Oxide Fuel Cells. Springer. London. 2009. pp.285-286.
[18] H. U. Anderson, F. Tietz. Interconnects in: S; C.Singhal, K. Kendall (eds), High temperature solid oxide fuel cell: fundamentals, design and applications. *Elsevier. Oxford.* 2003. pp. 174-175.
[19] K. Tanabe, K. Mismo, Y. Ono, H. Hattori. New solids acids and bases. Kodansha. Tokyo Elsevier. New York, 1989, pp. 41-47.
[20] C. Lahouse, A. Aboulayt, F. Maugé, J. Bachelier, J. C. Lavalley. Acidic and basic properties of zirconia—alumina and zirconia—titania mixed oxides. *J. Mol. Catal.* 84 (1993) 283-297.
[21] M. P. Rosynek, R. J. Koprowski, G. N. Dellisante. The nature of catalytic sites on lanthanum and neodymium oxides for dehydration/dehydrogenation of ethanol. *J. Catal.* 122 (1990) 80-94.

[22] C. Shi, A. B. Walters, M. A. Vannice. NO reduction by CH_4 in the presence of O_2 over La_2O_3 supported on Al_2O_3. *Appl. Catal. B: Environm.* 14 (1997) 175-188.
[23] S. J. Conway, J. A. Greig, G. M. Thomas. Comparison of lanthanum oxide and strontium-modified lanthanum oxide-catalysts for the oxidative coupling of methane. *Appl. Catal. A: Gen.* 86 (1992) 199-212.
[24] S. J. Korf, J. G. Van Ommen, J. R. H. Ross. The oxidative coupling of methane over Sm_2O_3 and La_2O_3. *Stud. Surf. Sci. Catal.* 67 (1991) 117-126.
[25] M. L. Pisarello, C. Saux, E. E. Miró, C. A. Querini. Stability of K-La_2O_3 catalysts during the combustion of Diesel soot. High frequency CO_2 pulses and soot-catalyst contacting studies. *Stud. Surf. Sci. Catal.* 139 (2001) 141-148.
[26] L.A. Isupova, V. A. Sadykov, L. P. Solovyova, M. P. Andrianova, V. P. Ivanov, G. N. Kryukova, V. N. Kolomiichuk, E. G. Avvakumov, I. A. Pauli, O. V. Adryushkova, V. A. Poluboyarov, A. Y. Rozovskii, V. F. Tretyakov. Monolith perovskite catalysts of honeycomb structure for fuel combustion. *Stud. Surf. Sci. Catal.* 91 (1995) 637-645.
[27] J. A. B. Bourzutschky, N. Homs, A. T. Bell. Conversion of synthesis gas over $LaMn_{1-x}Cu_xO_{3+\lambda}$ perovskites and related copper catalysts. *J. Catal.* 124 (1990) 52-72.
[28] H. Provendier, C. Petit, C. Estournes, A. Kiennemann. Dry reforming of methane. Interest of La-Ni-Fe solid solutions compared to $LaNiO_3$ and $LaFeO_3$. *Stud. Surf. Sci. Catal.* 119 (1998) 741-746.
[29] G. C. Araújo, S. M. de Lima, J. M. Assaf, M. A. Peña, J. L. G. Fierro, M. C. Rangel. Catalytic evaluation of perovskite-type oxide $LaNi_{1-x}Ru_xO_3$ in methane dry reforming. *Catal. Today* 133-35 (2008) 129-135.
[30] R. Pereñíguez, V. M. González-DelaCruz, J. P. Holgado, A. Caballero. Synthesis and characterization of a $LiNiO_3$ perovskite as precursors for methane reforming reactions catalysts. *Appl. Catal. B: Environm.* 93 (2010) 346-353.
[31] G. C. Araújo, S. M. Lima, M. C. Rangel, V. La Pagola, M. A. Peña, J. L. G. Fierro. Characterization of Precursors and Reactivity of $LaNi_{1-x}Co_xO_3$ for the Partial Oxidation of Methane. *Catal. Today* 107 (2005) 906-912.
[32] D. Andriamassinoro, R. Kieffer, A. Kiennemann, P. Poix. Preparation of stabilized copper-rare earth oxide catalyst for the synthesis of methanol from syngas. *Appl. Catal. A: Gen.* 106 (1993) 201-212.

[33] J. M. Grau, L. M. Gómez-Sainero, L. Daza, X. L. Seoane, A. Arcoya. Effect of barium and lanthanum oxides on the properties of Pt/KL catalysts in the n-heptane dehydrocyclization. *Stud. Surf. Sci. Catal.* 130 (2000) 2525-2530.
[34] S. Becker, M. Baerns. Oxidative coupling of methane over La_2O_3-CaO catalysts. Effect of bulk and surface properties on catalytic performance. *J. Catal.* 128 (1991) 512-519.
[35] E. E. Miró, F. Ravelli, M. A. Ulla, L. M. Comaglia, C. A. Querini. Catalytic diesel soot elimination on Co-K/La_2O_3 catalysts: Reaction mechanism and the effect of NO addition. *Stud. Surf. Sci. Catal.* 130 (2000) 731-736.
[36] P. Gronchi, E Centola, R. Del Rosso. Dry reforming of CH4 with Ni and Rh metal catalysts supported on SiO_2 and La_2O_3. *Appl. Catal. A: Gen.* 152 (1997) 83-92.
[37] Y. Dong, C. Jianshe, H. Qing, L. Kuiren. Effects of lanthanum addition on corrosion resistance of hot-dipped galvalume coating. *J. Rare Earths.* 27 (2009) 114-118.
[38] B. J. Hwang, Y. W. Tsai, G. T. K. Fey, J. F. Lee. Effect of lanthanum dopant on the structural and electrical properties of $LiCoVO_4$ cathode materials investigated by EXAFS. *J. Power Sources.* 97-98 (2001) 551-554.
[39] P. Ghosh, S. Mahanty, R. N. Basu. Lanthanum-doped $LiCoO_2$ cathode with high rate capability. *Electrochim. Acta* 54 (2009) 1654-1661.
[40] M. C. Rangel, C. L. Pieck, G. Pecchi, N. S. Fígoli, P. Reyes. Effect of the Solvent used during Preparation on the Properties of Pt/Al_2O_3 and Pt-Sn/Al_2O_3. *Ind. Eng.Chem. Res.* 40 (2001) 5557-5563.
[41] X. Chen, Y. Liu, G. Niu, Z. Yang, M. Bian, A. He, High temperature thermal stabilization of alumina modified by lanthanum species, *Appl. Catal. A:Gen.* 205 (2001) 159-172.
[42] B. Ersoy, V. Gunay, Effects of La_2O_3 addition on the thermal stability of γ-Al_2O_3 gels. *Ceram. Int.* 30 (2004) 163-170.
[43] L. J. Alvarez, J. P. Jacobs, J. F. Sanz, J. A. Odriozola, The thermostabilising effect of La doping on γ-Al_2O_3. A molecular dynamics simulation study. *Solid State Ionics.* 95 (1997) 73-79.
[44] Heck, R. M.; Farrauto, R. J.; Catalytic Air Pollution Control, Van Nostrand Reinhold: New York, 1995.
[45] M. C. Rangel, M. F. A. Carvalho. Impacto dos catalisadores automotivos no controle da qualidade do ar. *Quím. Nova.* 26 (2003) 265–277.

[46] H. Arai, M. Machida. Thermal stabilization of catalyst supports and their application to high-temperature catalytic combustion. *Appl. Catal. A: Gen.* 138 (1996) 161-176
[47] H. Schaper, E. B. M. Doesburg, P. H. M. De Korte, L. L. Van Reijen. Thermal stabilization of high surface area alumina. *Solid State Ionics.* 16 (1985) 261-265.
[48] P. Burtin, J.P. Brunelle, M. Pijolat, M. Soustelle. Influence of surface area and additives on the thermal stability of transition alumina catalyst supports. II: Kinetic model and interpretation. *Appl. Catal.* 34 (1987) 239-254.
[49] B. Béguin, E. Garbowski, M. Primet. Stabilization of alumina by addition of lanthanum. *Appl. Catal.* 75 (1991) 119-132.
[50] F. Oudet, P. Courtine, A. Vejux. Thermal stabilization of transition alumina by structural coherence with $LnAlO_3$ (Ln = La, Pr, Nd). *J. Catal.* 114 (1988) 112-120.
[51] M. Bettman, R. E. Chase, K. Otto, W. H. Weber. Dispersion studies on the system $La_2O_3/\gamma\text{-}Al_2O_3$. *J. Catal.* 117 (1989) 447-454.
[52] H. Arai, H. Fukuzawa. Research and development on high temperature catalytic combustion. *Catal. Today* 26 (1995) 217-221.
[53] D.L. Trimm. Catalytic combustion (review). *Appl. Catal.* 7 (1983) 249-282.
[54] R. A. D. Betta. Catalytic combustion gas turbine systems: the preferred technology for low emissions electric power production and co-generation. *Catal. Today* 35 (1997) 129-135.
[55] X. Jiang, R. Zhou, P. Pan, B. Zhu, X. Yuan, X. Zheng. Effect of the addition of La_2O_3 on TPR and TPD of $CuO/\gamma\text{-}Al_2O_3$ catalysts. *Appl. Catal. A: Gen.* 150 (1997) 131-141.
[56] C. F. Cullis, B. M. Willatt. Oxidation of methane over supported precious metal catalysts. *J. Catal.* 83 (1983) 267-285.
[57] T. Y. Chou, C. H. Leu, C. T. Yeh. Effects of the addition of lanthana on the thermal stability of alumina-supported palladium. *Catal. Today,* 26 (1995) 53-58.
[58] Z. Cheng, Q. Wu, J. Li, Q. Zhu. Effects of promoters and preparation procedures on reforming of methane with carbon dioxide over Ni/Al_2O_3 catalyst. *Catal. Today* 30 (1996) 147-155.
[59] Å. Slagtema, U. Olsbye, R. Blom, I. M. Dahl. The influence of rare earth oxides on Ni/Al_2O_3 catalysts during CO_2 reforming of methane. *Stud. Surf. Sci. Catal.* 107 (1997) 497-502.
[60] R. Blom, I. M. Dahl, Å. Slagtema, B. Sortland, A. Spjelkavika, E. Tangstad. Carbon dioxide reforming of methane over lanthanum-

modified catalysts in a fluidized-bed reactor. *Catal. Today* 21 (1994) 535-543.
[61] L. Cao, Y. Chen, W. Li. Effect of La_2O_3 added to NiO/Al_2O_3 catalyst on partial oxidation of methane to syngas. *Stud. Surf. Sci. Catal.* 107 (1997) 467-471.
[62] J. T. Richardson, B. Turk, M. V. Twigg. Reduction of model steam reforming catalysts: effect of oxide additives. Appl. Catal. A: Gen. 148 (1996) 97-112.
[63] S. Natesakhawat, O. Oktar, U. S. Ozkan. Effect of lanthanide promotion on catalytic performance of sol-gel Ni/Al_2O_3 catalysts in steam reforming of propane. *J. Mol. Catal. A: Chem.* 241 (2005) 133-146.
[64] S. Natesakhawat, R. B. Watson, X. Xueqin, U. S. Ozkan. Deactivation characteristics of lanthanide-promoted sol-gel Ni/Al_2O_3 catalysts in propane steam reforming. *J. Catal.* 234 (2005) 496-508.
[65] M. C. Alvarez-Galvan, R.M. Navarro, F. Rosa, Y. Briceño, F. Gordillo Alvarez, J.L.G. Fierro. Performance of La,Ce-modified alumina-supported Pt and Ni catalysts for the oxidative reforming of diesel hydrocarbons. *Int. J. Hydrogen Energy.* 33 (2008) 652–663.
[66] M. C. Alvarez-Galvan, R.M. Navarro, F. Rosa, Y. Briceño, M.A. Ridao, J.L.G. Fierro. Hydrogen production for fuel cell by oxidative reforming of diesel surrogate: Influence of ceria and/or lanthana over the activity of Pt/Al_2O_3 catalysts. *Fuel.* 87 (2008) 2502-2511.
[67] J. C. Amphlett, R. F. Mann, B. A. Peppley, P. R.Roberge, A. Rodrigues, J. P. Salvador. Simulation of a 250kW diesel fuel processor/PEM, fuel cell system. *J. Power Sources,* 71 (1998)179–184.
[68] L. Xiancai, H. Quanhong, Y. Yifeng, C. Juanrong, L. Zhihua. Effects of sol-gel method and lanthanum addition on catalytic performances of nickel-based catalysts for methane reforming with carbon dioxide. *J. Rare Earths,* 26 (2008) 864-868.
[69] R. Bouarab, O. Cherifi, A. Auroux. Effect of basicity created by La_2O_3 addition on the catalytic properties of $Co(O)/SiO_2$ in CH_4+CO_2 reaction. *Thermochim. Acta* 434 (2005) 69-73.
[70] L. Zhang, W Li, J. Liu, C. Guo, Y. Wang, J. Zhang. Ethanol steam reforming reactions over $Al_2O_3.SiO_2$-supported Ni–La catalysts. *Fuel.* 88 (2009) 511–518.
[71] T. Miki, T. Ogawa, M. Haneda, N. Kakuta, A. Ueno, S. Tateishi, S. Matsuura, M. Sato. Enhanced oxygen storage capacity of cerium

oxides in CeO$_2$/La$_2$O$_3$/Al$_2$O$_3$ containing precious metals. *J. Phys. Chem.* 94 (1990) 6464-6467.

[72] A. Trovarelli, C. de Leitenburg, M. Boaro, G. Dolcetti. The utilization of ceria in industrial catalysis. *Catal. Today.* 50 (1999) 353-367.

[73] F. Deganello, A. Longo, A. Martorana. EXAFS study of ceria–lanthana-based TWC promoters prepared by sol–gel routes. *J. Solid State Chem.* 175 (2003) 289–298.

[74] F. Deganello, A. Martorana. Phase analysis and oxygen storage capacity of ceria-lanthana-based TWC promoters prepared by sol-gel routes. *J. Solid State Chem.* 163 (2002) 527-533.

[75] N.E. Bogdanchikova, S. Fuentes, M. Avalos-Borja, M.H. Farías, A. Boronin, G. Díaz. Structural properties of Pd catalysts supported on Al$_2$O$_3$-La$_2$O$_3$ prepared by sol-gel method. *Appl. Catal. B: Environm.* 17 (1998) 221-231.

[76] T. Chen, H. Chen, H. Chen, L. Wang. J. Shi, Effect of La on the thermal stability of Pd/γ-Al$_2$O$_3$ catalytic membranes, *Ceram. Int.* 27 (2001) 883-887.

[77] M. I. Litter. Heterogeneous photocatalysis. Transition metal ions in photocatalytic systems. *Appl. Catal. B: Environm.* 23 (1999) 89–114.

[78] M. Uzunova-Bujnova, R. Todorovska, D. Dimitrov, D. Todorovsky, Lanthanide-doped titanium dioxide layers as photocatalysts, *Appl. Surf. Sci.* 254 (2008) 7296–7302.

[79] D. Xu, L. Feng, A. Lei, Characterizations of lanthanum trivalent ions/TiO2 nanopowders catalysis prepared by plasma spray, *Journal of Colloid and Interface Science.* 329 (2009) 395–403.

[80] M.R. Hoffmann, S.T. Martin, W. Choi, D.W. Bahnemann. Environmental applications of semiconductor photocatalysis. *Chem. Rev.* 95 (1995) 69-96.

[81] K.T. Ranjit, I. Willner, S.H. Bossmann, A.M. Braun. Lanthanide oxide doped titanium dioxide photocatalysts: effective photocatalysts for the enhanced degradation of salicylic acid and t-cinnamic acid. *J. Catal.* 204 (2001) 305-313.

[82] H.R. Kim, T.G. Lee, Y.-G. Shul. Photoluminescence of La/Ti mixed oxides prepared using sol–gel process and their pCBA photodecomposition. *J. Photochem. Photobiol. A: Chem.* 185 (2007) 156-160.

[83] J. Liqiang, S. Xiaojun, X. Baifu, W. Baiqi, C. Weimin, Fu Honggang. The preparation and characterization of La doped TiO$_2$ nanoparticles and their photocatalytic activity. *J. Solid State Chem.* 177 (2004) 3375-3382.

[84] J. Lin, J. C. Yu. An investigation on photocatalytic activities of mixed TiO,-rare earth oxides for the oxidation of acetone in air. *J. Photochem. Photobiol. A: Chem.* 116 (1998) 63-67.
[85] Y. Huang, Y. Xie, L. Fan, Y. Li, Y. Wei, J. Lin, J. Wu, Synthesis and photochemical properties of La-doped $HCa_2Nb_3O_{10}$, *Int. J. Hydrogen Energy*. 33 (2008) 6432–6438.
[86] A. A. Nada, M.H. Barakat, H.A. Hamed, N.R. Mohamed, T.N.Veziroglu. Studies on the photocatalytic hydrogen production using suspended modified TiO_2 photocatalysts. *Int. J. Hydrogen Energy*. 30 (2005) 687-691.
[87] H. Kato, K. Asakura, A. Kudo, Highly efficient water splitting into H_2 and O_2 over lanthanum-doped $NaTaO_3$ photocatalysts with high crystallinity and surface nanostructure. *J. Am. Chem. Soc.* 125 (2003) 3082-3089.
[88] J. S. Wang, S. Yin, M. Komatsu, T. Sato T. Lanthanum and nitrogen co-doped $SrTiO_3$ powders as visible light sensitive photocatalyst. *J. Eur. Ceram. Soc.* 25 (2005) 3207-3212.
[89] M. S. Ramos, M, de S. Santos, L. P. Gomes, A. Albornoz, M. C. Rangel. The influence of dopants on the catalytic activity of hematite in the ethylbenzene dehydrogenation. *Appl. Catal. A: Gen.* 341 (2008) 12-17.
[90] M. de S. Santos, A. Albornoz, M. C. Rangel. The Influence of the Preparation Method on the Catalytic Properties of Lanthanum-doped Hematite in the Ethylbenzene Dehydrogenation. *Stud. Surf. Sci. Catal.* 162 (2006)753-760.
[91] M. de S. Santos, S. G. Marchetti, A. Albornoz, M. C. Rangel. Effect of the preparation method on the properties of hematite-based catalysts with lanthanum for styrene production. *Stud. Surf. Sci. Catal.* (submitted).
[92] F. Cavani, F. Trifiró. Alternative processes for the production of styrene. *Appl. Catal. A: Gen.* 133 (1995) 219-239.
[93] E. H. Lee. Iron oxide catalysts for dehydrogenation of ethylbenzene in the presence of steam. *Catal. Rev.* 8 (1973) 285-305.
[94] A. Schüle, U. Nieken, O. Shekhah, W. Ranke, R. Schlögl, G. Kolios. Styrene synthesis over iron oxide catalysts: from single crystal model system to real catalysts. *Phys. Chem. Chem. Phys.* 9 (2007) 3619-3634.
[95] A. Kotarba, I. Kru, Z. Sojka. Energetics of potassium loss from styrene catalyst model components: reassignment of K storage and release phases. *J. Catal.* 211 (2002) 265-272.

[96] M.Muhler, R. Schlögl, A. Reller, G. Ertl. The nature of the active phase of the Fe/K-catalyst for dehydrogenation of ethylbenzene. *Catal. Lett.* 2 (1989) 201-210.
[97] G. R. Meima, P. G. Menon. Catalyst deactivation phenomena in styrene production. *Appl. Catal. A: Gen.* 212 (2001) 239-245.
[98] J. Matsui, T. Sodesawa, F. Nozaki. Activity decay of potassium-promoted iron oxide catalyst for dehydrogenation of ethylbenzene. *Appl. Catal.* 51 (1989) 203-211.
[99] I. Rossetti, E. Bencini, L. Tretini, L. Forni. Study of the deactivation of a commercial catalyst for ethylbenzene dehydrogenation to styrene. *Appl. Catal. A: Gen.* 292 (2005) 118-123.
[100] L. Huerta, A. Meyer, E. Choren. Synthesis, characterization and catalytic application for ethylbenzene dehydrogenation of an iron pillared clay. *Micropor. Mesopor. Mat.* 57 (2003) 219-227.
[101] A. C. Oliveira, A. Valentini, P. S. S. Nobre, J. L. G. Fierro, M. C. Rangel. Non Toxic Fe-Based Catalysts for Styrene Synthesis. *The Effect of Salt Precursor and Aluminum Promoter on the Catalytic Properties. Catal. Today*, 85 (2003) 49-57.
[102] W. P Addiego, W. Liu, T. Boger. Iron oxide-based honeycomb catalysts for the dehydrogenation of ethylbenzene to styrene. *Catal. Today.* 69 (2001) 25-31.
[103] G. A. El-Shobaky, N. R. E. Radwan, M. S. El-Shall, A.M. Turky, H.M.A. Hassan. Physicochemical, surface and catalytic properties of nanocrystalline CuO–NiO system as being influenced by doping with La_2O_3. Colloids and Surfaces A: *Physicochem. Eng. Aspects.* 345 (2009) 147–154.
[104] J. Yu, Guisheng, D. Mao, G. Lu. Effect of La_2O_3 on catalytic performance of Au/TiO_2 for CO oxidation. *Acta Physico-Chimica Sinica*, 24 (2008) 1751-1755.
[105] Y. Xu, X. Zhang, H. Li, Y. Qi, G. Lu, S. Li. Promotion effect of lanthanum addition on the catalytic activity of zirconia supported platinum and tungstophosphoric acid catalyst for n-pentane isomerization. *Appl. Surf. Sci.* 255 (2009) 6504–6507.
[106] S. Damyanova, L. Daza, J. L. G. Fierro. Surface and catalytic properties of lanthanum-promoted Ni/sepiolite catalysts for styrene dehydrogenation. *J. Catal.* 159 (1996) 150-161.
[107] Z. Wangcheng, L. Guanzhong, G. Yanglong, G. Yun, W. Yunsong. Synthesis of Ln-doped MCM-41 mesoporous materials and their catalytic performance in oxidation of styrene. *J. Rare Earths*, 26 (2008) 59-65.

INDEX

A

abatement, 12
absorption, 15
acetone, 8, 14, 34
acid, vii, 11, 14, 23, 33, 35
acidity, 5, 10, 25
acrylonitrile, 15
activation energy, 23
active site, 15, 17, 19
additives, 5, 31, 32
adsorption, 10
alcohols, 13
aldehydes, 13
amines, 13
ammonium, 16, 17, 18, 19, 20, 21, 22
anatase, 14
atoms, 23

B

band gap, 15
barium, 30
basicity, 11, 17, 32
batteries, 5
beneficial effect, 10, 25
benzene, 18, 21
boilers, vii, 5, 9
butadiene, 15

butadiene-styrene, 15

C

Ca^{2+}, 1, 6, 15
candidates, 1
carbon, vii, 3, 10, 12, 16, 22, 23, 31, 32
carbon dioxide, 10, 31, 32
carbon monoxide, vii, 3, 12, 22, 23
catalysis, 1, 5, 25, 33
catalyst, 9, 10, 11, 12, 13, 14, 16, 17, 18, 19, 21, 23, 29, 31, 32, 34, 35
catalyst deactivation, 9, 10
catalytic activity, 3, 9, 11, 19, 22, 23, 34, 35
catalytic properties, 3, 22, 32, 35
catalytic reaction, 13
cathode materials, 5, 30
ceramic, 1, 6, 13, 27
cerium, 11, 32
character, 3
chemical stability, 13
chromium, 16
CO2, 29, 31, 32
cobalt, 7, 11, 12
coherence, 31
coke, vii, 7, 9, 10, 11, 12, 16, 17, 21, 25
coke formation, 10, 16
combustion, vii, 3, 5, 6, 7, 9, 29, 31
competition, 12

compounds, vii, 1, 3, 12, 15, 16, 23, 25
conduction, 15
conductivity, 1, 28
configuration, 13
copper, vii, 9, 12, 22, 29
corrosion, 5, 30
cost, 9, 11
crystal growth, 12
crystal structure, 15
crystalline, 5, 22
crystallinity, 34
crystallites, 11, 23
crystals, 18

D

decay, 9, 35
decomposition, 3, 27
defects, 5, 14
degradation, 13, 14, 33
degree of crystallinity, 22
dehydration, 3, 28
dehydrocyclization, 3, 30
deposition, vii, 9, 10, 11, 16, 17, 25
desorption, 9
detoxification, 13
diesel fuel, 32
diffusion, 6
dispersion, 5, 9, 10, 11, 25
dissociation, 10
dopants, vii, 12, 13, 16, 17, 22, 25, 34
doping, 13, 14, 22, 30, 35

E

electrical conductivity, 5
electrical properties, 30
electrolyte, 1, 28
electron, 13, 14, 15, 17
electrons, 15
emission, 22
endothermic, 15
equilibrium, 15

ethanol, 3, 7, 12, 28
EXAFS, 12, 30, 33
extraction, 9

F

ferric ion, 20
filters, 1
filtration, 13
flame, 9

G

gasification, 16
gel, 33

H

heptane, 3, 30
heterogeneous catalysis, 7
host, 15
hydrocarbons, 8, 9, 10, 11, 13, 32
hydrogen, 9, 10, 11, 12, 13, 14, 15, 34
hydrogenation, vii, 3, 8, 23
hydrolysis, 19
hydroxide, 14, 16, 17, 18, 19, 20, 21, 22
hydroxyl, 14
hydroxyl groups, 14
hypothesis, 12

I

impregnation, 10
impurities, 16
inhibition, 14
interface, 6, 17
ion-exchange, 15
ions, 1, 13, 17, 18, 19, 21, 33
iron, vii, 16, 17, 19, 21, 22, 34, 35
isolation, 23
isomerization, vii, 8, 23, 35

Index

L

lanthanide, 11, 13, 32
lanthanum, vii, 1, 3, 5, 6, 7, 9, 10, 11, 12, 13, 14, 15, 16, 17, 18, 19, 20, 21, 22, 23, 25, 27, 28, 29, 30, 31, 32, 33, 34, 35
lattices, 23
low temperatures, 22

M

magnesium, 1
magnetic properties, 1, 27
membranes, 6, 8, 13, 33
mesoporous materials, 35
metal nanoparticles, 15
metal oxides, 10, 17
metals, 5, 9, 12, 33
methanol, 3, 29
Mg^{2+}, 1, 6
migration, 16
mixing, 16, 19, 20, 21
model system, 34
molar ratios, 17
molecular dynamics, 30
molecules, 14
monolayer, 11

N

nanoparticles, 33
natural gas, 10
Nd, 31
neodymium, 22, 28
nickel, vii, 7, 10, 11, 12, 22, 23, 32
nitrogen, 34
noble metals, 12

O

organic compounds, 13

oxidation, vii, 7, 8, 9, 10, 12, 14, 16, 22, 23, 32, 34, 35
oxygen, 7, 9, 11, 12, 14, 17, 22, 32, 33

P

palladium, 9, 13, 31
performance, 17, 30, 32, 35
permission, iv
perovskite oxide, 1
phase transformation, 5, 14
phase transitions, 6
phenol, 8, 14
photocatalysis, 33
photocatalysts, 15, 33, 34
photodegradation, 8
photoluminescence, 14
platinum, vii, 11, 15, 23, 35
pollution, 12
polymeric membranes, 13
polymers, 15
polystyrene, 15
potassium, 16, 19, 20, 21, 22, 27, 34, 35
probability, 23
promoter, 1, 10, 15, 16
propane, 7, 11, 32
properties, vii, 1, 3, 11, 13, 14, 15, 16, 17, 19, 25, 27, 28, 30, 33, 34
pyrolysis, 27

R

radicals, 14
radius, 14, 17
reactants, 16, 19, 20, 21
reactions, vii, 3, 7, 25, 29, 32
reactivity, 13, 15, 23
recombination, 14
requirements, 1
resistance, vii, 5, 10, 11, 12, 17, 19, 20, 21, 25, 30
room temperature, 16
rutile, 14

S

selectivity, vii, 7, 8, 10, 12, 19, 21, 22, 25
semiconductor, 13, 33
shape, 15
silica, vii, 11, 12
simulation, 30
sintering, 5, 7, 9, 12, 13, 17, 18, 19
SiO_2, 7, 12, 30, 32
sodium, 18, 19, 21, 22
sodium hydroxide, 18, 19, 22
sol-gel, 17, 19, 22, 32, 33
solid oxide fuel cells, 1, 25, 28
solid solutions, 1, 29
solid state, 1
species, 6, 10, 12, 15, 17, 19, 20, 21, 22, 30
specific surface, vii, 5, 7, 11, 16, 17, 18, 19, 23
spectroscopy, 12, 18, 20
stabilization, 6, 11, 12, 21, 30, 31
stabilizers, 6
steel, 5
storage, 7, 12, 32, 33, 34
strategy, 16
strong interaction, 6
strontium, 1, 29
styrene, vii, 8, 15, 16, 17, 19, 21, 23, 34, 35
sulfur, 11
superconductor, 27
surface area, 5, 9, 11, 13, 16, 17, 18, 19, 31
surface energy, 6
surface layer, 6
surface properties, 30
synthesis, 3, 15, 29, 34

T

temperature, vii, 7, 10, 12, 13, 23, 28, 30, 31
thermal stability, 5, 6, 7, 8, 9, 12, 13, 22, 25, 30, 31, 33
thin films, 27
titania, vii, 13, 14, 23, 28
titanium, 13, 14, 33
toluene, 18, 21
toxicity, 16
TPA, 8, 23
transformation, 5, 12, 13, 14
transition metal, 13

U

ultraviolet irradiation, 15
UV irradiation, 14

V

vacancies, 11, 12, 14
valence, 1, 15
versatility, 25

W

wavelengths, 15

X

XPS, 17

Y

yttrium, 12

Z

zinc, 12
zirconia, vii, 1, 23, 28, 35
zirconium, 22